THE ZEPPELIN

The downing of *LZ 37* by aviation artist Charles H. Hubbell (1899-1971). The *LZ 37* was based in Gontrode, Belgium. *LZ 37* was part of a raid with Zeppelin *LZ 38* and *LZ 39*. While returning, she was intercepted in the air by Reginald Warneford in his Morane Parasol during its first raid on Calais on 7 June 1915. Warneford dropped six 20 pounds (9.1 kg) Hales bombs on the zeppelin which caught fire and crashed into the convent school of Sint-Amandsberg, next to Ghent, Belgium, killing two nuns.

THE ZEPPELIN
–AN ILLUSTRATED HISTORY–

PHIL CARRADICE

FONTHILL

An early French airship/balloon photographed in 1904 (a colour autochrome) by Léon Gimpel (1873-1948).

Fonthill Media Limited
Fonthill Media LLC
www.fonthillmedia.com
office@fonthillmedia.com

First published in the United Kingdom and the United States of America 2017

British Library Cataloguing in Publication Data:
A catalogue record for this book is available from the British Library

Copyright © Phil Carradice 2017

ISBN 978-1-78155-505-7

The right of Phil Carradice to be identified as the author of this work has been asserted by him in accordance with the Copyright, Designs and Patents Act 1988.

All rights reserved. No part of this publication may be reproduced, stored in a retrieval system or transmitted in any form or by any means, electronic, mechanical, photocopying, recording or otherwise, without prior permission in writing from Fonthill Media Limited

Typeset in 10.5pt on 13pt Sabon LT Std
Printed and bound by CPI Group (UK) Ltd, Croydon, CR0 4YY

CONTENTS

	Introduction	7
1.	Balloons and Airships: The Early Days	9
2.	Zeppelins Prepare for War	17
3.	The Zeppelin War	20
4.	Post-War	62
	Bibliography	96

German Dirigible *Hansa* arriving at Potsdam Harbor *c.* 1915.

Ferdinand Adolf Heinrich August Graf von Zeppelin *c.* 1915.

INTRODUCTION

This book does not claim to be a full and complete history of Zeppelins and airships. While, on the one hand, it is merely a collection of photographs with captions, those evocative images and informed text still make it a book that tries to catch the atmosphere of the Zeppelin age and the people who lived in it.

For a brief period in the early twentieth century, it really seemed as if the future of air travel lay with the giant airships of Count von Zeppelin, Hugo Eckener, and the rest. The First World War ended that dream, fixed-wing aircraft superseding the slow-moving and unwieldy airships. As weapons of war the Zeppelins were never truly successful, although they did manage to terrify huge numbers of unknowing and naïve civilians—perhaps more by imagination than by any practical manifestation of their power.

The Zeppelin crews of the First World War spent hours in the air, cold and hungry—very little thought was ever given to their creature comforts. They flew with the ever-present prospect of a horrendous death, either by fire or by falling thousands of feet to the ground. As vehicles of mass destruction, the Zeppelins were remarkably ineffective. Their real value, if people had only realised it, lay in their ability to make silent reconnaissance missions over enemy territory and sea lanes.

It was only in the post-war days that the public began to realise that airships offered a form of air travel that was comfortable, mostly stable, and, sometimes, even luxurious. The *Graf Zeppelin* and the *Hindenburg* were the height of elegance, comparable to the luxury liners of the day and with considerably more opportunity for sight-seeing. Unfortunately, they had two major defects—they were vulnerable to the elements and, due to the hydrogen that kept them aloft, they were also highly flammable.

The *Hindenburg* disaster of 1937 effectively spelled the end of the giant airship as a viable commercial enterprise, but for almost half a century these wonderful machines had cruised elegantly through the clouds. Now, most people will see airships only at events like golf tournaments where the Goodyear Blimp regularly cruises over the fairways to take photographs.

It was not always like this. There was a time that Zeppelins and airships in general were state-of-the-art machines. They were of their age and era, and as such they deserve to be remembered.

Hans Rudolf Schulze (1870-1951) produced many patriotic postcards in the early years of the First World War.

CHAPTER ONE

BALLOONS AND AIRSHIPS: THE EARLY DAYS

One of the greatest challenges ever faced by mankind was undoubtedly the conquest of the air. The ability to soar like a bird through the skies was a dream that obsessed scientists and romantic adventurers for hundreds, if not thousands, of years. It was a dream that was to come to fruition in the technologically vibrant nineteenth century.

The dream, of course, had begun long before. In Italy, during the renaissance, the painter and inventor Leonardo da Vinci had made surprisingly effective and intriguing sketches for a flying machine. However, the technology to turn those images into reality was not available and da Vinci's ideas remained just a fanciful piece of early science fiction.

In Medieval Europe, legends of men who had actually flown were commonplace. The mythological story of Daedalus, who built gliders for himself and his son Icarus, fashioning the wings out of feathers and wax, was a talisman to many—despite the tragedy that befell Icarus who flew too close to the sun and so melted the wax of his wings.

The philosopher and writer Francis Bacon thought that the answer to man's quest for flight lay in a globe that could be filled with liquid fire, while Bishop Wilkins believed that a flying machine might be driven by steam. French locksmith Besnier is believed by many to have made several flights, albeit of short duration, using a pair of oars to propel him through the air.

It was all fanciful stuff, highly dubious to say the least, until the advent of the Montgolfier brothers. These two French inventors began to experiment with paper bags powered by heat and on 5 June 1783, at Annonay, they launched a cloth and wooden-framed balloon filled with hot air; the balloon shot into the air like a rocket to land 1½ miles away, where it was promptly attacked by superstitious villagers who saw it as a demon and hacked the structure to pieces.

Another Montgolfier experiment, using hydrogen instead of hot air, ended in disaster when the gas expanded at altitude and destroyed the canister that held it. Clearly the world was not yet ready for such technological advances as the use of hydrogen. However, one thing was clear—at this stage, at least—the future of aviation lay in balloons, not fixed-wing, heavier-than-air machines.

A few months after the Montgolfiers failed hydrogen flight, on 21 November 1783, Jean Philatre de Rozier and the Marquis D'Arlandes became the first men to actually fly in a hot air balloon. Desperate to maintain an adequate supply of hot air for their journey the two Frenchmen even installed a small furnace in their craft, regularly stoking it with lengths of wood. Inevitably, the fabric of their balloon caught fire several times, the flames having to be extinguished with a wet sponge, and the intrepid aviators had no idea where they might end up. In fact they flew for nearly five miles before coming down in safety.

Inspired by this feat, numerous balloon flights now took place—including several using hydrogen to fill the balloon bag—and it was not long before ballooning was being regarded as a new and adventurous sport. As ever, people wanted to set records, to achieve fame and be remembered by posterity. In keeping with this aim, in the late 1830s Charles Green and John La Mountain undertook incredible long-distance flights of 500 and 800 miles respectively and made themselves household names in the process.

Inevitably, military minds began to consider the advantages of flight. For reconnaissance of enemy positions, balloons were an ideal vehicle, observers being able to see far more clearly—and for greater distances—than men on horseback. There was also the decided benefit of using them to direct artillery fire; something that was done successfully, although on a small scale, by both Union and Confederate forces during the American Civil War.

The age of offensive aerial warfare had already arrived some years earlier when, in 1849, the Austrian army officer Franz Ucharius planned and oversaw a bombing attack on the Italian city of Venice. Horrified and frightened almost out of their wits by what was happening, hundreds of Venetians stood in wonder as bombs rained down on their ancient city. It was, as H. G. Wells might have said 'the shape of things to come'.

During the Siege of Paris in the Franco-Prussian War of 1870–71, balloons were used to maintain communications between the beleaguered city and the outside world. Most were quickly shot down by Prussian artillery, but several were able to make their way to safety. By some miracle, one even managed to reach the northern part of Norway, a journey of well over 1,000 miles, before it was brought down by the vagaries of the weather.

Balloons, of course, had a major disadvantage, which severely limited their use for military purposes—they could not be steered. Once launched, their progress, or not, depended largely on the strength and direction of the wind. It was therefore imperative that some means of steering balloons through the air should be found. In the second half of the nineteenth century, the drive for steerable balloons—dirigibles, as they were known—began to dominate the thinking of those who were determined to take safely to the air.

A hot air balloon, designed and built by the Montgolfier brothers. Such contraptions were beyond the understanding of the French peasants who attacked one of the Montgolfier's creations as it descended from the sky, thinking that it had come from the Devil. Despite such superstition, the Montgolfier's work was instrumental in the development of ballooning.

Henri Giffard had invented a steam-driven airship in 1852, but the engine was so heavy that the experiment ended in failure. It was only when the petrol engine came into use that airship development was able to move on to the next stage.

In 1897 German inventor Dr Wolfert built a cigar-shaped balloon that was powered by a small petrol engine and although his first flight ended in tragedy and disaster, it did not stop the Brazilian Alberto Santos-Dumont designing a similar machine. In 1901, he won the Deutsche de la Meurthe Prize after steering his dirigible from St Cloud, around the Eiffel Tower and back to St Cloud once again.

The Santos Dumont Air-Ship rounding the Eiffel Tower, on October 19th 1901.

Brazilian balloonist and aviator, Alberto Santos-Dumont, steers his dirigible around the Eiffel Tower to win the Deutsch de la Meurthe prize, on 19 October 1901. The journey from Parc St Cloud to the Eiffel Tower and back was supposed to be completed in under thirty minutes. Santos-Dumont over-ran the time limit, but as there had been a delay in picking up his mooring line he was allowed to claim the prize. It was a seminal moment in airship development and made Santos-Dumont into a national figure, both in Europe and back in his native Brazil. Sadly, he became increasingly depressed by the use of aeroplanes and airships in war and committed suicide, hanging himself on 23 July 1932.

VON ZEPPELIN TAKES TO THE SKIES

The achievement of Santos-Dumont firmly established the importance and the power of airships in the minds of the general public. And the man whose name now became synonymous with airships and air power was the German Count Ferdinand Zeppelin.

Count Ferdinand Adolf Heinrich von Zeppelin was born into an aristocratic family in 1838, long before the creation of modern Germany. Like most aristocratic young men from Germanic families at this time, he enlisted in the army, serving his home state of Württemberg. His career blossomed and he was sent to America to serve as an official observer with the Union forces during the American Civil War. This was when he became interested in ballooning and made his first ascent in an airship at Saint Paul in Minnesota.

During the Franco-Prussian War of 1870–71, von Zeppelin conducted himself with great aplomb and achieved national recognition and fame for his bravery. In 1890, aged sixty-two, he retired from the army and began to devote his considerable energies to the development of what he conceived as the weapon of the future—the dirigible airship.

Von Zeppelin's first submission and set of designs was made to the Prussian Airship Service in 1894. Although they initially agreed funding for developments, the Committee of the Airship Service soon changed its mind, deeming the proposals impractical, and withdrew its offer. Zeppelin was undaunted, but decided that the only way his design—and any future developments of that design—were ever going to become reality was for him to fund the enterprise himself.

Zeppelin's design was, in a way, quite simple. His airship would be constructed from a rigid aluminium frame, covered with an envelope of silk. Inside were a number of smaller gasbags, tiny 'balloonets' that kept the airship aloft. If one of them should happen to be punctured, the others would keep the airship flying, in much the same way as the sequence of watertight compartments on the *Titanic* were designed to keep the ship afloat.

Along with Philip Holzmann and others, Zeppelin now formed Gesellschaft zur Förderung der Luftschiffahrt—the Society for the Promotion of Airships. Zeppelin himself provided 441,000 marks for the enterprise, over half the working capital, and, at the huge floating hangar on Lake Mansell, work began on *LZ 1*.

When the *LZ 1* was tested in 1901 she reached a speed of 18 mph and the German public, desperate for success and recognition, took Count Zeppelin and his dirigible to their hearts.

Graf von Zeppelin was the person most obviously connected with airship design and development, his name being synonymous with dirigibles, but he was already an old man of sixty-nine when Santos-Dumont rounded the Eiffel Tower to win the Deutsche de la Meurthe prize. His energy and technical knowledge, however, put a great number of younger men to shame.

Von Zeppelin and Hugo Eckener in the gondola of Zeppelin *LZ 10* on 26 September 1907. Eckener was a journalist who was so impressed by the Count's skill and energy that he agreed to become a part-time publicist for the Zeppelin Company. He gained his licence to fly Zeppelins in 1911 and went on to take charge of the company after Zeppelin's death. In the years after the First World War he was the Captain of the *Graf Zeppelin* on many of her record-breaking voyages.

In the Kaiser's pre-war Germany, with the emphasis on all things martial and military, it was inevitable that the armed forces would become interested in von Zeppelin's airships. This shows the Count in the company of Kaiser Wilhelm and his entourage—the topic of conversation would, inevitably, have been airships and their value in war.

PRE-WAR DEVELOPMENTS

The first decade of the twentieth century saw both triumph and disaster for Count von Zeppelin. His company went into liquidation in December 1900 and it needed a public appeal to fund a new airship. *LZ 1* was succeeded by *LZ 2* and *3*, and the German army announced that it would purchase one of the airships, provided it had the ability to remain in the air for twenty-four hours and have a range of 700 km. In order to meet these requirements a new airship was needed and *LZ 4* was promptly built.

When the new airship was destroyed by the wind in the summer of 1908, it was public support that saved the day. Over 6 million marks were raised—thanks, partly, to the efforts of the journalist Dr Hugo Eckener—and *LZ 5* set out on a marathon thirty-seven-hour flight that so impressed the German General Staff that the airship, along with the earlier *LZ 3*, was promptly bought by the Army.

The DELAG Company (standing for Deutsche Luftschiffahrts-Aktiengesellschaft) was founded by Count von Zeppelin in late 1909. It was the world's first commercial airline and, despite the fact that several of the company's airships were accidentally destroyed, by the summer of 1912 the *LZ 10* (*Schwaben*, as she was known) had made over 200 flights, carrying 1,500 passengers on pleasure and business cruises. Zeppelin fever continued to grow.

It was the size of the giant airships that amazed the public. To see such gigantic machines take to the skies was a sight not to be missed. In keeping with their majestic appearance, the first airships were housed and flew from large bodies of water, like Lake Constance.

Both the Army and the Navy now realised, perhaps rather belatedly, that there was a future for military airships and bought several of von Zeppelin's machines. Even the loss of two of their first airships could not dampen the Navy's new-found enthusiasm.

The Zeppelin Company was not the only producer of rigid airships in Germany at this time. The Schutte-Lanz Airship Company had been founded by Johann Schutte of Danzig and other entrepreneurs in 1909, their vessels being formed and kept in shape by laminated plywood, rather than the aluminium girders used by Zeppelin. Marketed as being lighter, faster, and more durable than the Zeppelin airships, the Schutte-Lanz ships had a particular appeal to the Army, which carefully fostered the interest of this rival company to Count von Zeppelin.

It was not just the German public that was entranced by airships; everyone was fascinated by these huge flying machines. This rather romanticised early postcard view shows Zeppelin *LZ 3* in the skies above a German city in the happy and heady pre-war days. Thousands would have stopped to watch the progress of Count Zeppelin's marvellous machine and many more, who might never actually get to see an airship in real life, would have bought the postcard.

CHAPTER TWO

ZEPPELINS PREPARE FOR WAR

The years leading up to the outbreak of the First World War were dominated by two interlinked and dynamic emotions. Firstly came the German desire for offensive weapons—on land, on sea, and in the air—to rival and exceed those of Britain. And secondly, there was the incessant, frantic, and almost paranoid British fear of invasion.

Germany had come late to the world stage, the country being welded together by Otto von Bismarck from a series of large and small Germanic states in the final few years of the 1870s. Under the guidance, influence, and control of Prussian militarism, the German state, the German people, and the German Kaiser demanded their place on the world stage.

Germany's assets in the late-nineteenth and early twentieth centuries were formidable. They included a navy to rival Britain's, colonies in Africa and the South Seas, a fearsome military machine that had the will and the training to defeat almost any army in the world, and a dynamic industrial power base that rapidly propelled the country into a position of economic excellence.

The intense nationalism engendered by Kaiser Wilhelm II was best seen in the growth of the Navy and, in particular, the High Seas Fleet, but the enthusiasm of the German public for Count Zeppelin's airships almost matched it. These were, people felt, the ideal weapons to bring Britain (always Germany's most intractable foe) to her knees. They were terror weapons whose power would only be really appreciated once war began.

In Britain, during the first dozen years of the twentieth century, there was something of a 'Zeppelin mania'. It was part of a nation-wide paranoia and mass hysteria, with dozens of articles and books written about the invasion plans of the Kaiser and his demonic German hoards. One of the most controversial and influential of these was *The Invasion of 1910* by William Le Queux, a sensationalist novel that quickly became a best seller, and has now totally disappeared from the annals of British literature.

Erskine Childers' *Riddle of the Sands* was a much more substantial offering, one that has stood the test of time. Even Arthur Conan Doyle got in on the act, pitching Sherlock Holmes against German spy Von Bork. It was not one of Conan Doyle's most memorable stories, but it did give his great detective one of the best and most memorable lines in the whole of the Holmes canon: 'There's an east wind coming—such a wind as never blew on England yet; a good many of us may wither before its blast'.

Giant aerial armadas were part of the hysteria. Most people had never seen an airship and terror of the unknown was far greater than the commonplace. Fear of sudden and unexpected assault by Germany's massive Zeppelin fleet grew to demonic proportions, everyone expecting that, when and if hostilities between Britain and Germany did actually break out, there would be an immediate mass assault from the skies.

In fact Germany did not possess the giant Zeppelin fleet that everyone in Britain supposed. In August 1914, on the outbreak of the long-expected war, only seven airships were in service with the army. The navy had even less. Following the loss of *LZ 14* and *LZ 18* in 1913, they had only the *LZ 24* in service. Of the army's seven airships, three were stationed in the east and only four in the west.

It is hardly surprising then, that the dramatic assault from the skies on British towns and cities did not materialise. That would come later, when the Zeppelin fleets had been enlarged and the Kaiser had given his permission to attack strategic targets in Britain.

LZ 11 can be seen here in its hangar. Originally in service with the DELAG company, when war broke out *LZ 11* it was immediately taken over by the German army. Unfortunately, the airship was accidentally destroyed while being towed into its hangar on 1 October 1915.

The outside of *LZ 11*'s passenger cabin is shown here, the photograph having been taken in the days before she was adopted by the military. Water, which acted as ballast, is being released prior to take off. At this stage airship travel was still a great novelty.

LZ 13 first took to the air on 3 May 1913, operating as one of the DELAG airships. In just over a year, before the winds of war stopped commercial flying, the *LZ 13* carried nearly 10,000 passengers and made 419 flights. By the time war broke out, however, she was already obsolete and was soon scrapped.

CHAPTER THREE

THE ZEPPELIN WAR

For the men flying the Zeppelins (officers and petty officers, every one of them) the war started badly. In the first month of the conflict, in the days before the elaborate trench systems were created in northern Europe, and in the mistaken belief that airships could carry out tactical daylight raids against specific military targets, three of the western based Zeppelins attempted to bomb French positions that were holding up the German advance. All three of the giant airships were shot down by anti-aircraft fire.

On the eastern front, a raid on the railway yards at the Russian town of Mlawa at the end of August 1914 also saw another of the Zeppelins destroyed. It was clearly apparent that Zeppelins could not be used to bomb military targets (tactical targets as they might be termed), at least not in the day time.

The difference between tactical and strategic targets, in the context of Zeppelin attacks, needs to be outlined. It was a 'nice' difference, but one that was to become very important in the months and years ahead.

Attacking tactical targets—a line of advancing soldiers, a squadron of cavalry, or even a battleship at sea—spelled clear and apparent danger for the Zeppelins. Their speed was too slow, either to hit and run away or simply to manoeuvre around the potential targets, all of which had the ability to fire back. The primitive nature of bombing and bomb aiming in particular at that time meant that the airships would have to operate at very low level, where their bulk made them easy marks for gunners on the ground.

Strategic targets such as railway stations, marshalling yards, factories, and dockyards (instillations that were not only influential in the war effort, but also had the added benefit of being stationary) offered a much more lucrative reward, provided the airships could attack in sufficient numbers. And also, of course, provided they were shielded by the cover of darkness.

Right: When Europe tumbled, almost unsuspecting, into war, Germany's initial attack in August 1914 was, courtesy of the Schlieffen Plan, aimed at France. The Kaiser and all of the German General Staff knew, however, that the country's most dangerous opponent was Britain—as this 1914 postcard shows. Suddenly, thanks to the Zeppelins, Britain was no longer safe behind the English Channel.

Below: LZ 23 was one of the ill-fated airships that were misguidedly assigned to attack French positions on 21 August 1914. Operating at just a few hundred feet, the slow-moving Zeppelin provided an easy target for the French gunners and thousands of bullets were pumped into the airship. It was a hard lesson to learn, but Zeppelins were clearly not designed for low level attacks on well-defended military positions.

Out of control, *LZ 23* came down between the French and German armies near Bandonvillers. The crew attempted to burn the remains of the airship before the French could arrive, but failed as there was not enough gas left in the ruptured tanks. They were taken prisoner by the jubilant French artillery men.

The *LZ 17*, another airship taken over by the German Army when war was declared, made an attack on the port of Antwerp while it was being defended by British and Belgian troops in August 1914. She spent twelve hours aloft, depositing over 800 kg of bombs onto the city, before turning and flying back to her base. The *LZ 17* was taken out of service in 1916.

The first Zeppelin raid on London took place on the night of 31 May 1915 when dozens of bombs were dropped from *LZ 38*. Seven people died in the attack and thirty-five were injured. *LZ 38* was one of the new 'P Class' of Naval Zeppelins. Soon after this initial raid, on 7 June, the *LZ 38* was bombed and destroyed in its hangar in Belgium—an attack by the Royal Naval Air Service that caused the German High Command to rethink the positioning of its Zeppelin bases. As a consequence, both the German Army and Navy withdrew their Zeppelins from Belgium.

The first aerial victory by an aeroplane over a Zeppelin occurred on 7 June 1915 when Sub Lt Reginald Warneford located LZ 37 over Ostend. From a height of 11,000 feet, Warneford dropped six 10 lb bombs onto the Zeppelin and watched in amazement as they struck home. The blast turned his aircraft upside down and he was lucky to escape with his life. He was forced to land in enemy territory for emergency repairs to his engine before taking off and heading home. Meanwhile, the *LZ 37* fell in a ball of smoke and flame.

Warneford was awarded the Victoria Cross for his action in bringing down the *LZ 37*. Not a great pilot—his instructor had said that he would either achieve great things or kill himself—Warneford died just a few weeks after his victory, in a flying accident. This 'In Memorium' card was produced shortly after his death.

1915

After the disasters of the first month of the war, it was not until the beginning of 1915 that Germany's much lauded fleet of airships finally came into operational existence. Even then, their effect was limited.

The Kaiser was, at first, reluctant to agree to indiscriminate raids on Britain. It would, he said, be contravening Article 25 of the 1907 Hague Convention, which expressly forbade aerial attacks on centres of population. In the end it was public pressure as much as the advice of his generals and admirals that forced the Kaiser to unleash his airships.

Before the war, Ernst Lehmann had commanded the commercial dirigible *Sachsen* on passenger flights between Leipzig and Saxony. He went on to command Zeppelins throughout the war and, later, to become commanding officer of the giant *Hindenburg*. As a wartime Zeppelin captain he saw and understood the desire of the German people to destroy British morale. He later wrote:

> The German people hoped for and expected extraordinary accomplishments from the airships. Their animosity was directed principally against England who had laid a hunger blockade while she remained impregnable from land and sea. Consequently, the cry of German public opinion for an attack from the air was doubly vociferous.
>
> *Zeppelin: The Story of Lighter-Than-Air Craft*, Ernst Lehmann

The first Zeppelin raid on Britain took place on the night of 19–20 January 1915. Two navy airships, the *LZ 24* and *LZ 27*, crossed the North Sea and dropped their bombs on Great Yarmouth and several of the surrounding villages. Four civilians were killed and almost twenty were injured by shrapnel and flying debris. It was a small enough attack, but it was a seminal moment; it proved to the world that Britain was no longer safe behind the protecting wall of the English Channel.

Over the following months several more attacks were made, most of them on targets in southern England. The first raid on London took place on the night of 31 May–1 June. It was carried out by *LZ 38*, one of the new 'P Class' of naval Zeppelins, and was commanded by Eric Linnarz. Seven people were killed by the bombs, thirty-five being injured or wounded.

Linnarz was the man who, during an earlier attack on Southend, had hurled a message from the gondola of *LZ 38*. It was a chilling, almost demonic communication and it summed up the attitude of the Zeppelin crews at this time: 'You English. We have come and will come again soon. Kill or cure. German'.

Further attacks on Ramsgate, Southend, and London (again) saw *LZ 38* alone drop 8,360 kgs of bombs, but, on 17 May, Linnarz and his crew were lucky to escape with their lives. In what was to be the first attack from the air on a Zeppelin, Flight Lieutenant Bigsworth tracked *LZ 38* to its base near Ostend and dropped a 20 lb bomb onto its fabric. One crewman was killed, the outer fabric slashed and the port engine put out of action, but the Zeppelin managed to limp home.

The Zeppelin attacks spread a wave of terror across Britain, particularly in dense centres of population like London. The image of a silent airship nosing through the clouds, seemingly impervious to all attempts to bring it down, was the stuff of nightmares, not just for children, but for logical, thinking adults. Ernst Lehmann's description of the effect of the raids may not be too far from the truth:

> Frightened out of their sleep, millions of people fled to the cellars, stumbling through the darkness and cowering together at every detonation. No-one knew where the next bomb would strike, and this uncertainty was nerve wracking. Women broke out in hysterical crying, children wept and men, no longer able to endure it down below, rushed out into the streets with balled fists, as if wanting to strangle an enemy hovering invisible in the clouds above.
>
> *Zeppelin: The Story of Lighter-Than-Air Craft*, Ernst Lehmann

In fact the Zeppelins were not as indestructible as the British public thought. Several of the giant airships had been already brought down by anti-aircraft fire; in April 1915, Linnarz's *LZ 38* was destroyed in its hangar, when Lanoe Hawker, Britain's first air ace, attacked the airship sheds at Gontrode. Hawker's attack was one of great daring and skill, his weapons consisting of just a few small bombs and some home-made hand grenades.

On the night of 6–7 June, Flight Sub-Lieutenant Reginald Warneford became the first pilot to destroy a Zeppelin in the air when a bomb from his Morane-Saulnier Scout smashed into *LZ 37* as she was approaching her base in Belgium. The airship fell in a ball of flame, much of the debris dropping onto a convent—two of the Sisters were killed, but one lucky Zeppelin crewman had a miraculous escape when he fell from the blazing airship and landed on a bed recently vacated by one of the nuns.

Airship raids were suspended for the summer of 1915, as the short nights and good visibility deprived the Zeppelins of the invisibility that was so important to their success. The raids recommenced in September, an attack on 7–8th of the month, depositing more than 4,800 kgs of bombs on London, Norwich, and Middlesbrough.

The targets were dock instillations and railway stations, but, inevitably, many of the missiles were wildly off-target and fell onto streets and houses. And so the terror continued. By the end of 1915, over thirty Zeppelin raids had been launched against England. With the defenders (RFC pilots and ground-based anti-aircraft weapons alike) seemingly unable to do much to stop them, there was every reason to suppose that 1916 would bring only greater problems.

A Zeppelin, raiding London, is caught in the searchlights, September 1915, a British postcard view. The message on the back of the card reads 'Dear Sister, we send you this photo of one of the Zeppelins that came over London. We had the pleasure of seeing it from our street. The Zep is on the top of the searchlights and the stars are from our guns'.

LZ 40 made five separate raids on England during 1915. As well as dropping over 9,000 kg of bombs on various targets, she also attacked and destroyed a British submarine. She was herself hit by lightning during a thunderstorm above Cuxhaven on 3 September 1915. The resulting crash killed nineteen crew members.

LZ 24 (originally identified as *L 3*) took part in the first raid on England during the night of 19 January 1915, when Great Yarmouth and the surrounding area were bombed. The ship was destroyed in an accident just one month later, making a forced landing in Denmark. The crew successfully abandoned ship and headed off into internment, but *LZ 24*, with her engines still running, was blown out to sea and lost.

One of the more successful Zeppelins, *LZ 36*, undertook seventy-four missions over the North Sea, scouting for British warships and merchant craft. She raided England four times, dropping nearly 6,000 kg of bombs and also attacked a submarine, which was severely damaged. On 16 September 1916, the Zeppelin was eventually destroyed by fire in its hangar at Fuhlsbüttel, north of Hamburg, when the *LZ 31* (which it was alongside) ignited while being inflated. The fire spread too quickly to get *LZ 36* out of the hangar.

Above: Fear of Zeppelin attacks remained strong, from 1914 through to the end of the war. It did not stop the British laughing at their fear, as this Punch cartoon from 7 October 1914 clearly shows. The cyclist in the cartoon might well be advised to douse his light, but a blackout was coming as fear of Zeppelin attacks grew stronger.

Left: The Zeppelin attacks horrified the British public and the press was quick to pick up on the emotion. 'Baby killers,' the Zeppelin crews were called. As far as the Germans were concerned the attacks were in retribution for the blockade that caused great hardship in the German cities. This *Punch* cartoon from 9 February 1916 neatly sums up both attitudes.

THE FLIGHT THAT FAILED.

The Emperor. "WHAT! NO BABES, SIRRAH?"
The Murderer. "ALAS! SIRE, NONE."
The Emperor. "WELL, THEN, NO BABES, NO IRON CROSSES."

[Exit murderer, discouraged.

This cartoon takes a more 'one sided' view of the Zeppelin raids—no babies killed means no Iron Crosses for the crew says the artist.

1916

In May 1916 the first of the 'Super Zeppelins' came into active service. These massive airships, capable of carrying a far greater payload of bombs than their earlier counterparts, were faster and had a greater range than anything that had yet taken to the skies. Unfortunately for them, their arrival coincided with improved defensive techniques by the British.

Aircraft were now equipped with a combination of ordinary explosive rounds and incendiary bullets. The idea was simple, the standard rounds would rip into the fabric and hull of any Zeppelin under attack and the incendiary bullets would then ignite the gas, which, it was hoped, was now escaping from the ruptured gas cells.

In the days before conscription, the British Government seized any and every opportunity to encourage men to join up, even using fear of Zeppelin attacks as a good reason to seek the security of the front lines. It seems an unlikely line of reasoning, even by 1914–1915 standards.

That is what happened on the night of 2–3 September, when Lieutenant William Leefe-Robinson shot down *SL 11* over Cuffley in Essex. It was the first time a Zeppelin had been brought down over British soil, and Leefe-Robinson used bullets rather than bombs to achieve his victory. Actually the *SL 11* was a wooden-framed Schutte-Lanz airship, not a Zeppelin, but the difference meant little to the public who immediately took Leefe-Robinson to their hearts or to the government, which promptly awarded the pilot a Victoria Cross.

Within a few weeks *LZ 72*, *LZ 74*, and *LZ 78* were all destroyed by British airmen—shooting down Zeppelins had become so 'commonplace' that the pilots, Frederick Sowrey and Wulstan Tempest, were simply awarded with a DSO each. *LZ 72* was the command of Heinrich Mathy, perhaps the most famous of all Zeppelin officers. He, along with the rest of his crew, died leaping from their blazing Zeppelin over Potters Bar.

By the end of the year six airships had been lost while attempting to bomb targets in England. Count Zeppelin was, apparently, so concerned about the disasters that, in a private discussion, he told Field Marshal von Hindenburg that airships had outlived their usefulness and that future bombing raids on England might be more successful if undertaken by fixed-wing aircraft.

A contemporary German print, originally published in *Illustrierte Zeitung*, shows what conditions on board the Zeppelins during a raid on England must have been like. Here mechanics work on the engines while gunners keep a sharp look out for British fighters. Ten or twelve hours in the air, over hostile territory, would have shredded anyone's nerves.

The Zeppelin raids, along with bombardment from the sea by German battlecruisers, brought home the fact that in this war, for the first time, the whole of Britain was in a combat zone. This shows bomb damage after one Zeppelin raid.

This postcard shows the shooting down of a Zeppelin (possibly *LZ 29*) by a motorised unit of French artillery. The image might owe more to imagination than to reality, but the joy of all anti-aircraft gunners in bringing down a Zeppelin was both real and, to the Zeppelin crews at least, very dangerous.

Like rabbits caught in a poacher's trap, the slow-moving Zeppelins were very vulnerable once they were highlighted in the glow of a searchlight. Here the 'low down thing that plays the low down game' is clearly illuminated, the postcard producer making no comment on the fact that British aircraft were also attacking German towns and industrial centres in the same way.

Count Zeppelin continued to design and build airships as the war progressed. This shows him in his element, aloft in one of his giant creations.

LZ 19 was one of nine Zeppelins detailed to raid Britain on 1 February 1916. Commanded by Odo Lowe, the airship developed engine trouble and crashed in the sea where early the following morning she was discovered by the British Trawler *King Stephen*. The German crew were standing on top of their airship, but Captain William Martin refused to rescue them and steamed away, leaving the Germans to their fate. There were no survivors. The story only became known when a message in a bottle, thrown into the sea by one of the German airmen, was picked up. As a result of his actions, Martin was pilloried in some quarters, applauded in others.

The burnt out remains of a crashed Zeppelin, brought down by anti-aircraft fire, are shown on this French postcard. The wreckage tells its own story—there was little hope of survival if gunfire was accurate.

The Zeppelin War 35

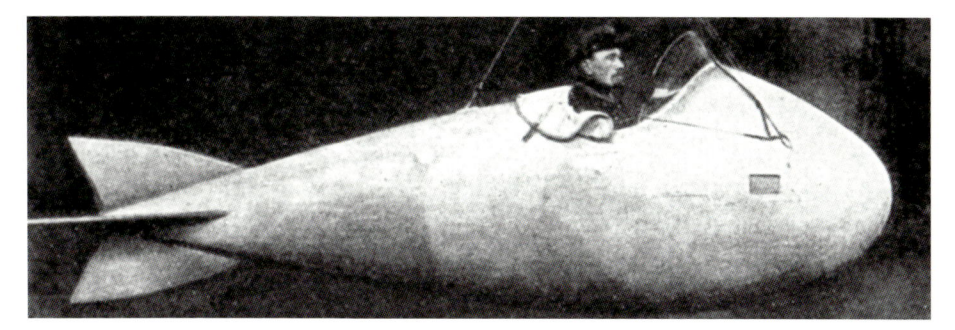

Height and good cloud cover were the Zeppelins best friends. But such things did present problems when it came to working out exactly where the airship was. One answer was to lower a small gondola or spahkorb from the main ship. It was a perilous task for the crewman in the contraption but it did at least give the Zeppelin crew an idea of their location.

A German drawing of men working in the forward compartment of an attacking Zeppelin. The compartment seems rather spacious and everything is well-ordered—artistic licence, perhaps, for the benefit of the German public.

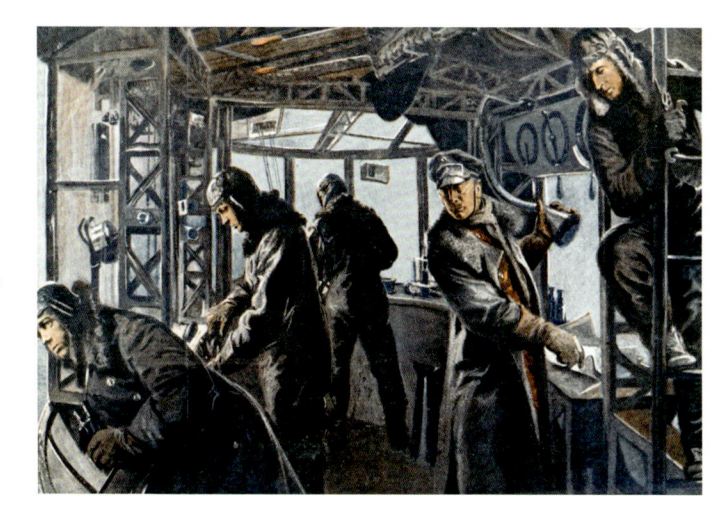

As the caption to this contemporary British postcard says, a Zeppelin is here 'Docked in the cradle of the deep'. Fear unleashed a wave of emotion, and hatred of the Zeppelin crews was commonplace.

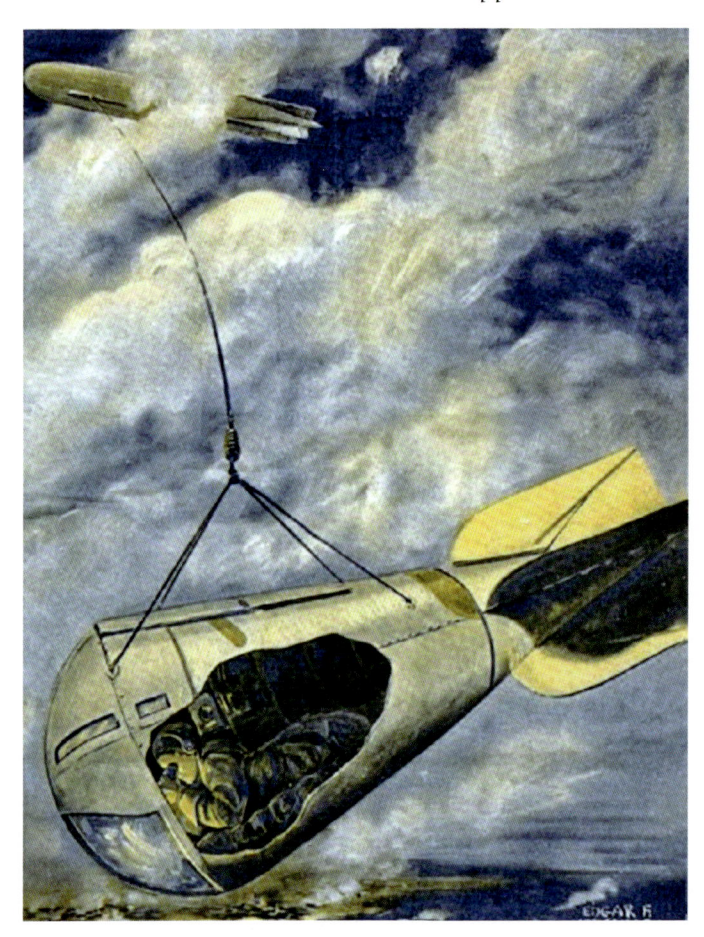

An observatory car suspended from a Zeppelin. *Scientific American* 23 December 1916.

On 31 January 1916, the Zeppelin *LZ 45* sailed on across the Midlands to the western coast of Britain. From here she made important radio transmissions regarding British shipping. It was one of forty-five reconnaissance flights made by *LZ 45* as well as a further fifteen bombing raids. *LZ 45* was taken out of commission in the spring of 1917.

Bad weather and bad luck prevented Zeppelins playing too big a part in the Battle of Jutland, which took place in May 1916, the only time that the two battle fleets of Britain and Germany came into contact during the whole war. But ten of them were there, high in the sky, observing the movements of Jellicoe and his ships.

LZ 65, which was destroyed by French anti-aircraft fire on 21 February 1916. The Zeppelin was attempting to bomb Vitry-le-François.

The Zeppelin

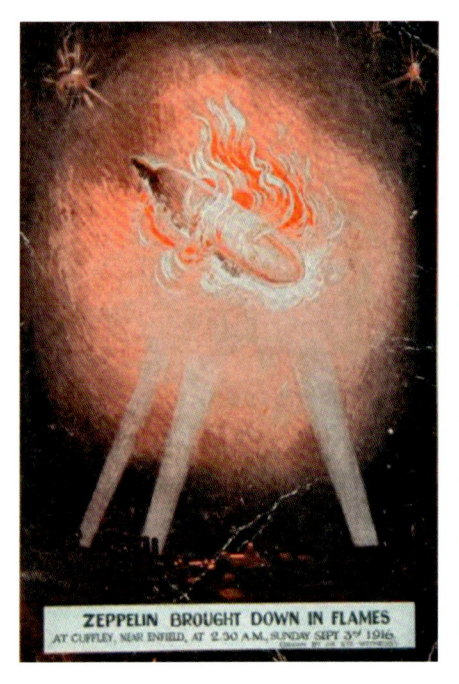

ZEPPELIN BROUGHT DOWN IN FLAMES
AT CUFFLEY, NEAR ENFIELD, AT 2.30 A.M., SUNDAY SEPT 3rd 1916.

The first airship to be shot down over British soil was not actually a Zeppelin, but, rather, the wooden-framed Schutte-Lanz *SL 11*. In the eyes of the public such differences were mere technicalities. On the night on 2–3 September 1916, Lieutenant William Leefe-Robinson, flying a BE2c aircraft and armed with incendiary bullets, intercepted the airship over Cuffley in Essex. After firing several drums of ammunition into the body of *SL 11*, Leefe-Robinson was overjoyed to see flames coming from the side of the raider. *SL 11* exploded and crashed, much to the joy of thousands who had gathered on the ground to watch the contest.

THREE GALLANT AIRMEN, ROBINSON, V.C., TEMPEST, AND SOWREY D.S.O.
Passed for publication by the Press Bureau on Oct 5th 1916.

Leef-Robinson was awarded the Victoria Cross three days after bringing down *SL 11*. He became a national hero as the man who, single-handedly, had saved the country from the perils of the German Zeppelins. This postcard from the time shows him (left) with two other 'Zeppelin killers,' Frederick Sowery and William Tempest. By the time Sowery and Tempest achieved their victories shooting down Zeppelins was not such a novelty and they were awarded only a DSO each.

In contrast to other airmen, whose deeds were rarely publicised, Leefe-Robinson was given the benefit of huge publicity by the British Propaganda Department. Awarded £2,000 by the city of Newcastle, he could not walk down the road without being accosted by hordes of young women. Unfortunately, he was later shot down by Baron Manfred von Richthofen and spent the final year of the conflict as a prisoner of war. Because of his fame, and the notoriety of his deed in shooting down *SL 11*, he was given harsh treatment in the camp. It ruined his constitution and he died in the flu epidemic of 1918–1919.

LZ 72, previously given the nomenclature *L 31*, was commanded by the most famous of all Zeppelin commanders, Kapitänleutnant Heinrich Mathy. A brave and dedicated Zeppelin pilot, Mathy once attacked a flotilla of British submarines, sinking one. His luck was not to last. The *LZ 72* was intercepted by Lt William Tempest and destroyed over Potters Bar on 2 October 1916. When the airship caught fire Mathy and his entire crew died.

Left: Rather than take his luck in the flames, Heinrich Mathy either leapt or fell from the blazing airship. Certainly his body was found well away from the wreckage and the remains of the other airmen. Mathy could hardly have hoped to survive the jump. The force of the fall caused his body to be partly embedded in the earth and although it was reported that he might have lived for a few moments, he died before he could be given medical attention.

Below: Despite the loss of so many airships, the Zeppelin raids continued throughout 1915 and 1916. This shows damage from a Zeppelin raid on Hull.

1917

Despite von Zeppelin's views, Korvettenkapitan Peter Strasser maintained his faith in the airships. He had been appointed commander of the Naval Airship Division in 1913 after Freidrich Metzing had been drowned when his Zeppelin came down in the sea; three years later, on 28 November 1916, he became the overall commander of German airships. Under his aggressive and determined direction the Zeppelin raids against England continued.

Strasser's enthusiasm for attacking England was matched by people such as Grand Admiral Tirpitz, the guiding genius behind the creation of the High Seas Fleet. Tirpitz was clear about the effect air raids had on London, believing that single bombs that, perhaps, killed one old woman or man were both ineffective and frightful. Massed assaults, however, were something totally different:

> If one could set fire to London in thirty places, then what in a small way is odious would retire before something fine and wonderful. All that flies and creeps should be concentrated on that city.
>
> Quoted in *The First Blitz*, Andrew Hyde

Peter Strasser believed that the best defence for the Zeppelins was to attack from a greater height where they would be above the relatively low ceiling of aircraft at the time. Unfortunately, greater height meant slower speeds, several of the airships having to have engines removed in order to reduce weight. This was a stopgap until, what the British pilots called, the new 'height climber' Zeppelins could be introduced.

The new high altitude Zeppelins attacked England for the first time on 16–17 March 1917. The raid was largely ineffective, but only one airship was lost out of a fleet of five; this was mainly due to the weather, rather than RFC attacks. Driven out over the sea by the wind, *LZ 86* suffered engine failure and, out of control, drifted over France where she was hit by anti-aircraft fire before crashing in flames close to Compiègne.

Perhaps the most successful raid of 1917 took place on the night of 23–24 May when bombs fell on an ammunition depot in Ramsgate. The size of the raiding airship fleets steadily increased, one raid in September consisting of nine vessels. An eleven airship raid on 19–20 October ended in disaster when no fewer than five airships were destroyed. This time it was a gale rather than British attacks that brought down the Zeppelins.

The story of the *LZ 50*, just one of the five lost airships, was a tragic one. She was blown southwards, out of control, over the Alps, the crew hoping to make Switzerland and safety. Having been in the air at high altitudes for over twenty-four hours, the oxygen supply on board was exhausted and many of the crew members were already unconscious. The airship smashed into a mountain peak and the control car gondola was ripped from its mountings before the airship rose once more into the air. The men in the gondola were the lucky ones. They survived while the mechanics in the motor gondolas were borne away. The ship struck another peak before disappearing out over the Mediterranean. She was never seen again.

One of the great legends of the Zeppelin war occurred in the autumn of 1917. In an attempt to relieve the hard-pressed General Lettow-Vorbeck, who was fighting a bitter and long-running campaign against British, South African, and Portuguese forces in what was then known as German East Africa, the *LZ 104* left Bulgaria at the end of November 1917. With a crew of twenty-two she was under the command of Ludwig Bockholt, an experienced Zeppelin flyer who early in the war had captured an English fishing boat while commanding the *LZ 23* and went on to take charge of two other airships. Known as the *Afrikaschiff*, the *LZ 104* reached the Sudan before she picked up a radio message telling her that Lettow-Vorbeck had surrendered.

The cartoon cover of *Punch* magazine 27 June 1917, featuring a Zeppelin.

The airship, which had never been intended to make a return journey, promptly turned around and flew back to her base at Yambol. It had been an amazing achievement, the *Afrikaschiff* having flown 6,800 km and been aloft for more than ninety hours. In as far as she was too late to help Lettow-Vorbek, the mission might have been a failure, but her achievement showed that airships were more than capable of undertaking long-range flights in all manner of weather conditions. In the years ahead it was to be an important factor.

So many Zeppelin men died in blazing airships that parachutes, then in their infancy, were issued to the crews. Ernst Lehmann, later commander of the *Hindenburg*, was clear about why these potentially lifesaving devices were ignored: 'Since the additional weight was at the expense of fuel and projectiles, we quickly abandoned them ... and continued to leave the decision of life or death to our own skill and luck'. Observers in balloons over the Western Front were not so cavalier and regularly used parachutes to escape their burning balloons.

After the initial disasters of August 1914, it had quickly become apparent that darkness and the cover of night were the best defence for the slow-moving airships. Once they were caught in the beams of searchlights, it was difficult for the Zeppelins to escape and their only hope was that British aircraft did not arrive on the scene before they could outrange the searchlights.

The use of airships was not confined to German forces. The Royal Naval Air Service regularly used them for anti-submarine patrols in the North Sea and out in the Western Approaches. Several were attached to the Dover Patrol. Early British dirigibles were basic, often just the fuselage of an aeroplane slung below the balloon envelope. With open cockpits, conditions out over the sea were harsh and draining for the pilots and observers.

OBSERVATION BALLOON OFF TO WATCH THE ENEMY.

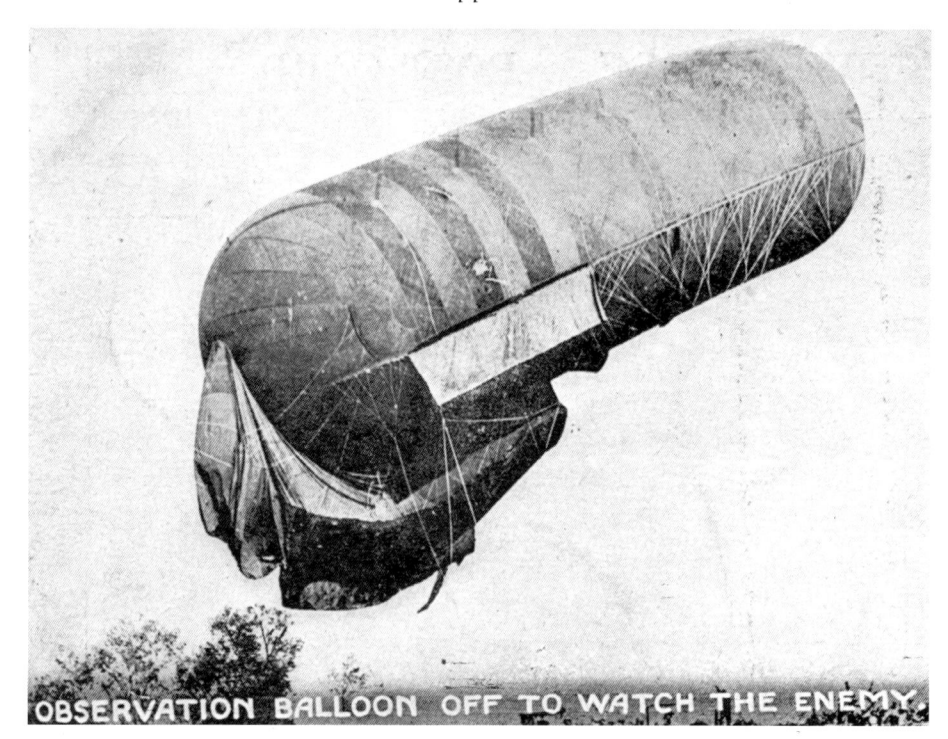

Observation balloons were also used by both sides during the war. Technically not airships, they were still well-used and were invaluable in passing back information on troop movements and, naturally, were often attacked by fighters. That was not the only hazard for the balloonists. Incorrectly filled balloons would lead to all of the gas gathering at one end of the envelope, hence the phrase 'Everything's gone pear shaped'.

Controls were fairly basic in the early Zeppelins, but things improved as newer airships were produced. This photograph, taken on board a Zeppelin in 1916, shows a panel of electrical apparatus used in bombing attacks.

A German newssheet shows the propaganda value of Zeppelins even if, by the end of 1916, there were many who questioned their real military value.

LZ 95 was part of a four-strong squadron of airships that attacked London on the night of 16–17 June 1917. She became separated from the other Zeppelins and, on the morning of 17th, was shot down by fighters over the sea near Great Yarmouth. There were three survivors, the rest of the crew were buried in Suffolk.

Coming down in the sea at least gave the Zeppelin crews some chance of survival, provided the descent was not too steep. Burning in the air meant almost certain destruction for both men and airship.

The crew of *LZ 39* are assembled in front of their airship, posing before a flight. The *LZ 39* was destroyed by French anti-aircraft fire on 17 March 1917 when many of the men shown here would have been killed or taken prisoner.

"The strafed Zepp." June 17. 17. No. 2. Photo: J.S. Maddell Leiston.

The sad, burned-out remains of a downed Zeppelin—the canvas has gone, just the skeleton-like ribs of the frame remains.

Forced to crash land near Bourbonne-les-Bains on 20 October 1917, *LZ 96* was captured virtually intact by the French soldiers who had brought her down. Allied airship builders seized on the chance to examine the ship and, consequently, her design was later hugely influential in the post-war development of British and American airships.

The Zeppelin War

A First World War public information poster.

By 1917, the heady pre-war days, and, indeed, the early months of the conflict when von Zeppelin was the darling of the Kaiser and the German establishment, were over. The Count, shown here in the company of the German Crown Prince watching one of his Zeppelins coming in to land, was an old man before war began and on 8 March 1917 he died, just prior to his eightieth birthday.

1918

Despite Strasser's enthusiasm, by the beginning of 1918 everyone knew that the day of the offensive airship had passed. The fixed-wing Gotha bombers continued to raid Britain and, thanks to Peter Strasser, so did the Zeppelins. The first Zeppelin attack of the year came on 12–13 March, but a thick cloud cover meant that the raid was largely ineffective.

A raid on 12–13 April saw a machine gunner on the *LZ 107* score several hits on an attacking aircraft, forcing it to crash land. The event was, possibly, the first occasion that gun fire from an airship had shot down an enemy plane. It was small consolation.

In a raid on London, launched on 5 August, Peter Strasser died when his airship was shot down over the sea. He had been the major advocate of the airship as an offensive weapon and he died in the last strategic raid of the war. It was almost the end of an era. Count von Zeppelin had died on 8 March the previous year and so did not live to witness the demise of his airships, but people like Ernst Lehmann were only too aware that the gigantic terror weapon of the early war years had finally been superseded:

> All in all, the role of the airship as a military weapon was over at this time. The development of air forces and defence batteries rendered it less and less useful over land, and it was replaced by the giant plane carrying heavy loads of bombs over the enemy positions at night.
>
> *Zeppelin: The Story of Lighter-Than-Air Craft*, Ernst Lehmann

The army dismantled its airships, keeping only four for reconnaissance duties. When a British Sopwith Camel, launched from a barge towed behind a destroyer, shot down *LZ 100* over the North Sea it was clear that a new phase of aerial war had arrived—the age of carrier-born aircraft. And that was a war in which Zeppelins had no place.

After von Zeppelin's death control of the airship company passed into the hands of Dr Hugo Eckener. He was astute enough to keep the company going for the rest of the war years and throughout the hard economic conditions of post-war Germany. A staunch anti-Nazi, Eckener was eventually blacklisted by Hitler's regime and lost his post. He died in 1954.

The Zeppelin War

Throughout the war years, *Punch* continued to publish cartoons using Zeppelins as a subject. This one from June 1917 takes a surreptitious 'dig' at Zeppelins, food rationing, and the German U-boat campaign, with a delighted citizen mistaking a little boy's model airship for a vegetable.

Above left: 'Ought we to grow up?' asks *Punch* in this cartoon from 23 February 1916.

Above right: A First World War cartoon postcard by Dudley Graham Buxton (1884-1951).

One of the greatest achievements of the Zeppelin war came in November 1917 when *LZ 104* set out to resupply German troops in East Africa. When she received news that the German forces had surrendered, the airship—originally intended to be broken up and the engines and canvas cannibalised for use on the ground—promptly turned around and flew back to her base in Bulgaria. The airship broke the long-distance record for a Zeppelin, flying 6,757 kilometres in just over ninety-five hours. The airship was later lost during a raid on Malta.

A view from the flight deck of *LZ 4*, originally published in *The Illustrated London News*.

The Zeppelin War

Right: Grand Admiral Tirpitz was one of the greatest advocates of air raids against Britain. Attacks by Zeppelins or by Gotha bombers, Tirpitz was clear that air raids were one of the best ways to destroy the morale of the British people.

Below: From the spring of 1917, Gotha bombers were increasingly used by the German air force, their air raids taking place alongside continued Zeppelin attacks. The bombers were hardy and massive machines and had the advantage of not being as vulnerable to the elements as the airships. This British drawing shows the dimensions of the aircraft and a list of its weapons and crew members.

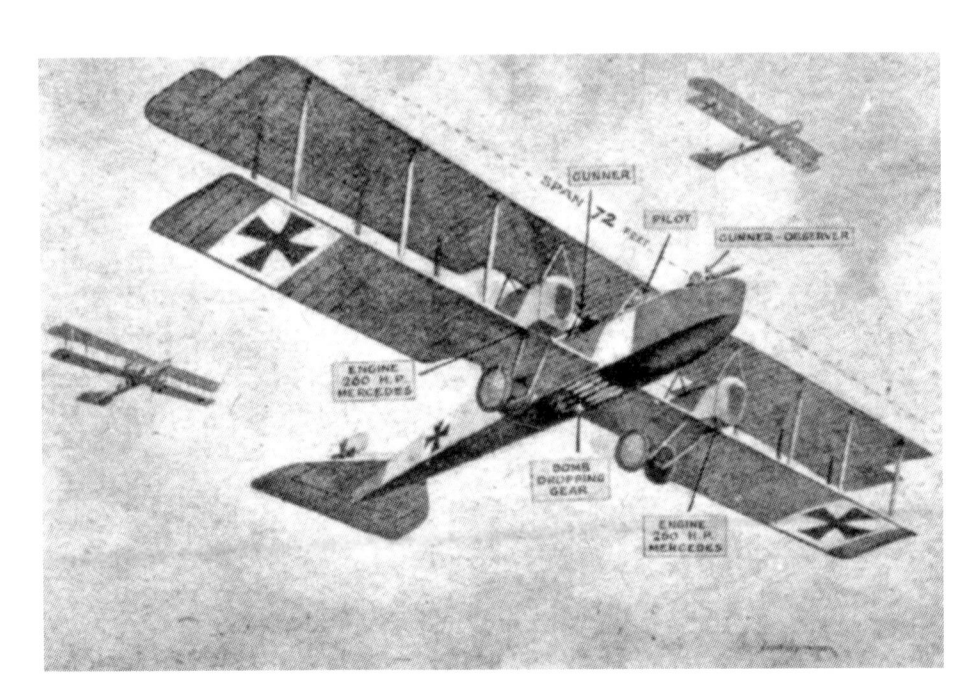

The loss of *LZ 112* on 5–6 August 1918 was a double blow for the German forces. On board the Zeppelin was Fregattenkapitän Peter Strasser, the overall commander of all German airships and a champion of continued airship attacks on Britain. *LZ 112* was intercepted and shot down by Major Egbert Cadbury and Captain Robert Leckie over the North Sea. There were no survivors.

Peter Strasser was a tireless advocate for Zeppelins. He had joined the Navy in 1876, rose steadily through the ranks and become a renowned gunnery officer before transferring to the airship division. As a naval officer he had the ear of Tirpitz, a fact that went some way towards the continuation of the Zeppelin raids on England, long after the airships had outlived their military usefulness.

Peter Strasser is pictured here alongside Count Zeppelin and Hugo Eckener. The photograph was taken in 1916, just two years later both Strasser and von Zeppelin would be dead.

On 5 January 1918, a sudden and devastating explosion tore through the Zeppelin hangars and warehouses at Ahlhorn, killing fifteen people and injuring over 130. The cause remained unknown, although sabotage was widely suspected. This photograph shows the wreckage of airships and hangars.

'Take away the night light, Mary,' says the young child in this *Punch* cartoon. 'I'd rather risk the dark than attract a Zeppelin'. The cartoon hides the fact that, because of air raids, a blackout had been introduced into Britain as early as 1916. It was nowhere near as total as the blackout imposed on the country in the Second World War, but fear of the Zeppelins led to this first lighting order being brought in and monitored by the police.

Above: *LZ 100* was the last airship destroyed during the First World War. On 11 August 1918, she was intercepted by Lt Culley, flying a Sopwith Camel, and shot down. The incident was a distinct nod to the future—Culley's Camel had been towed out to sea on a lighter or barge behind the destroyer HMS *Redoubt* and then launched to attack the Zeppelin. The age of the aircraft carrier, albeit in a primitive form, had arrived. It was an age in which, militarily at least, airships had no part.

Left: By the summer of 1918, Gotha bombers were faring no better than the Zeppelins. This postcard from the time shows the destruction of one of the raiders over Kent. With undisguised glee, the small poem at the top of the postcard reads 'Behold the end of a raiding Gotha/A prey to Kentish fire/Our boys at the guns have finished the Hun/And lit their funeral pyre'.

The Zeppelin War

Right: When the raiders, Zeppelins or Gothas, did get through and drop their bombs, the damage was significant. This photograph shows the effect of shrapnel from the bombs that has smashed windows and pock-marked the walls of a house.

Below: It was not just British cities that came under attack from the Zeppelins and, later, the Gotha bombers. This shows a huge bomb crater in a Paris street, the result of a Zeppelin raid in early 1917. Although Zeppelin pilots tried to target industrial areas such as factories and railways, the accuracy of their bombing was not always good and many residential and commercial areas were also hit.

This shows the nacelle of the airship *LZ 83*. Launched for the first time on 22 February 1917, over the next eighteen months she carried out fifteen reconnaissance missions on the Eastern Front and over the Baltic before undertaking three bombing missions against England. When the war ended, the *LZ 83* was handed over to France as part of the war reparations programme.

LZ 120 had begun life as a commercial airship with DELAG. Like other surviving Zeppelins, she became part of the reparations payments, being handed over to Italy in 1921. She duly arrived in Rome on Christmas Day 1921 and was renamed *Esperia*.

LZ 121 was yet another airship seized by the victorious Allies after the war. Originally intended for commercial flights to and from Sweden, she was gifted to the French and renamed *Mediterranee*.

Yet another airship handed over to the Allies, *LZ 114* was taken over by France on 9 July 1920. She claimed the world endurance record with a flight of 118 hours, but exploded off the coast of Italy in December 1923. Everyone on board died in the accident. This photograph was taken from the forward gondola of the airship.

A *Punch* cartoon bemoaning the evil of the Zeppelin attacks.

The Zeppelin

Even Bruce Bairnsfather, the most pre-eminent of all British First World War cartoonists, found the Zeppelins an easy target for his humour—as this example shows.

A Zeppelin flies high above the battleship *Seydlitz*, co-operation between the different arms of the German Navy continuing right to the end of the war.

CHAPTER FOUR

POST-WAR

The Zeppelin Company survived both the German defeat and the death of its founder. Dr Hugo Eckener took the lead, restarting DELAG in August 1919. Six of the airships still in service at the end of the war were handed over, as part of the reparations payments, to Britain, France, and Italy. Even the USA demanded that they be given an airship and the new *LZ 126*, built with German money by German workers, duly became the US Navy's *Los Angeles*. Dr Eckener himself delivered it across the Atlantic in October 1924.

The success of the *Los Angeles* established a firm working relationship between the US Navy and the Zeppelin Company. Arguably, that relationship was what saved the company from financial ruin in the hard economic conditions of 1920s Europe. The company built two helium-filled airships for the US Navy in 1931 and 1932, but, despite being excellent examples of their type, both of them were destroyed (by storm and technical failure) soon after they were commissioned.

In the 1920s Britain began to build airships with a view to using them for long-distance travel between England and outposts of the empire, such as India. The destruction of the *R 101*, which hit a hillside in France on 5 October 1930 while on her maiden voyage, was a body blow to the embryonic British airship industry. Of the fifty-four passengers and crew on board, forty-eight were killed and the tragedy spelled the end of Britain's brief flirtation with commercial airships.

The story of the *Graf Zeppelin* is well-known. Under the command of Eckener in 1929, this enormous airship made nearly 600 flights and was the first airship to circumnavigate the world. Her successor, the *Hindenburg*, flew for the first time in 1936 and was intended for Trans-Atlantic flights between Germany and the New World.

The spectacular destruction of the *Hindenburg* at Lakehurst, New Jersey, on 16 May 1937 killed thirty-seven people and started a controversy that has never ended. What caused the airship to suddenly explode and fall to ground in a ball of flame? Suggestions of sabotage, electrical strikes, even the flammability of the dope used to coat the airship's fabric have been made, but the answer has not been found. What is certain is the fact that it meant the end of hydrogen-filled airships.

Although the last of the Zeppelins, the *Graf Zeppelin II*, was used for reconnaissance of the British coast in the weeks before the Second World War broke out, there was no place for airships in the second great conflict of the twentieth century. The last Zeppelin, on the orders of Hermann Göring, was broken up.

Barrage balloons for defence and hot air balloons for leisure purposes soon became the limit for lighter-than-air craft. The future, it was clear, belonged to fixed-wing aeroplanes.

Post-war flying, a Zeppelin soars high over the bridge across the Rhine and the twin spires of Cologne Cathedral. Despite the military failures during the war, nobody doubted the commercial possibilities of airships—hence the glee with which the victorious nations accepted German Zeppelins as part payment of the war reparations.

The USA was not to be left out in the scramble for reparations and demanded that the Zeppelin Company supply them with an airship. They commandeered the *LZ 126* then being built in Germany, an act that caused some anger and resentment among the German people. Nevertheless, building *LZ 126* did keep many German workers in a job. This shows partly inflated gas cells in the belly of the airship.

Dr Eckener himself delivered the *LZ 126*, quickly renamed *Los Angeles* by the Americans, across the Atlantic in October 1924. The radio antennas of the airship are shown in this photograph.

This shows the radio room of *LZ 126 Los Angeles*. The airship was so successful that the newly formed Goodyear Zeppelin Company was commissioned to build two more airships for the US Navy.

Dr Eckener with Anton Witterman, the Navigator, on board the *LZ 126 Los Angeles* in 1924.

Above: LZ 126 leaving Friedrichshafen 12 October 1924.

Left: LZ 126 leaves the coast of Europe for America.

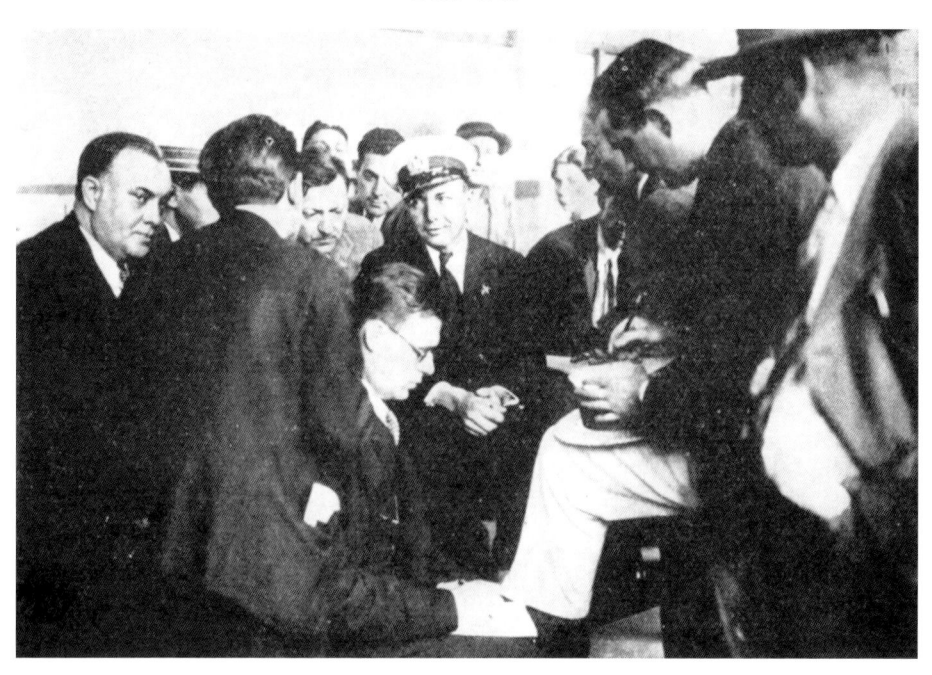

After the successful landing of *LZ 126*—which was renamed *Los Angeles* by the Americans—Ernst Lehmann besieged by the American press.

Ernst Lehmann, Hugo Eckener, Hans Flemming, President Calvin Coolidge and Navy Secretary Curtis Wilbur at the White House, 16 October 1924.

Based on the design of *LZ 96*, which had been captured virtually intact in 1917, the *Shenandoah*—'*Daughter of the Stars*'—was the first rigid airship built in the USA. She was a wonderful publicity machine for the US Navy, feted and stared at wherever she went. This photograph shows her under construction at Lakehurst in New Jersey.

On 3 September 1925, the *Shenandoah* ran into violent weather conditions while flying over southern Ohio. Out of control, the airship rose to 6,000 feet before plummeting back to earth. Tossed about like a rubber ball by the wind and pressure, she eventually slammed into the Ohio countryside. The airship was a total loss, fourteen of her crew dying in the crash.

Ernst Lehmann, Zeppelin commander of note and second-in-command of the *LZ 126* on her trip across the Atlantic—the first ever trans-Atlantic flight—in 1924.

The giant *Graf Zeppelin* under construction in Germany.

The *Graf Zeppelin* is shown here landing at Friedrichshafen. Under the command of Hugo Eckener she made 590 flights and was the first airship to circumnavigate the globe. During this amazing trip, no fewer than twelve full days were actually spent in the air.

Britain's airships had developed considerably since the days when they were little more than aircraft fuselages slung below a balloon, and in the post-war world they were considered ideal for making the fast, comfortable passage between Britain and her far-flung colonies. When built, the *R 101* was the largest airship in the world—a record she held until the *Hindenburg* arrived on the scene several years later. Sadly, the *R 101* crashed into a hillside in France on her maiden voyage on 5 October 1930. Forty-eight passengers and crew, including Lord Thomson, Britain's Secretary of State for Air, died in the crash.

Graf Zeppelin over Chicago, 28 August 1929. At the behest of American newspaper publisher William Randolph Hearst, whose media empire was the major commercial backer of the project with four staffers among the flight's nine passengers, the Graf's 'Round-the-World' flight commenced on 8 August 1929. It officially began and ended at Lakehurst Naval Air Station in New Jersey. This was a photo opportunity like no other.

Graf Zeppelin at Lakehurst 29 August 1929 at the end of the 'Round-the-World' flight.

Graf Zeppelin with a 1929 Packard at Lakehurst.

Graf Zeppelin draws a crowd at Friedrichshafen.

The controls of *LZ 127 Graf Zeppelin*.

On one Atlantic crossing a mission was undertaken to repair the fin and a small team of riggers set out across the top of the *Graf Zeppelin* in the late night storm to reattach the canvas and restore control to the captain. It was an extremely dangerous mission and it took place thousands of feet above the raging North Atlantic, during the repair process the *Graf Zeppelin*'s engines were turned off although this caused the airship to slowly descend, twice during the process the engines had to be powered up to push the ship back up to a safer altitude whilst the men on top held on for their lives. It is hardly surprising there is slight camera shake.

Graf Zeppelin flying over Cairo on 10 April 1931.

Graf Zeppelin over Jerusalem, 26 April 1931.

In 1935, while on a regular trip to Brazil *Graf Zeppelin* was forced to cruise around aloft for four days and was running short of food. They telegraphed a German ship, the SS *España*, and managed to pick up provisions.

Repairing the *Graf Zeppelin* over the South Atlantic in 1934.

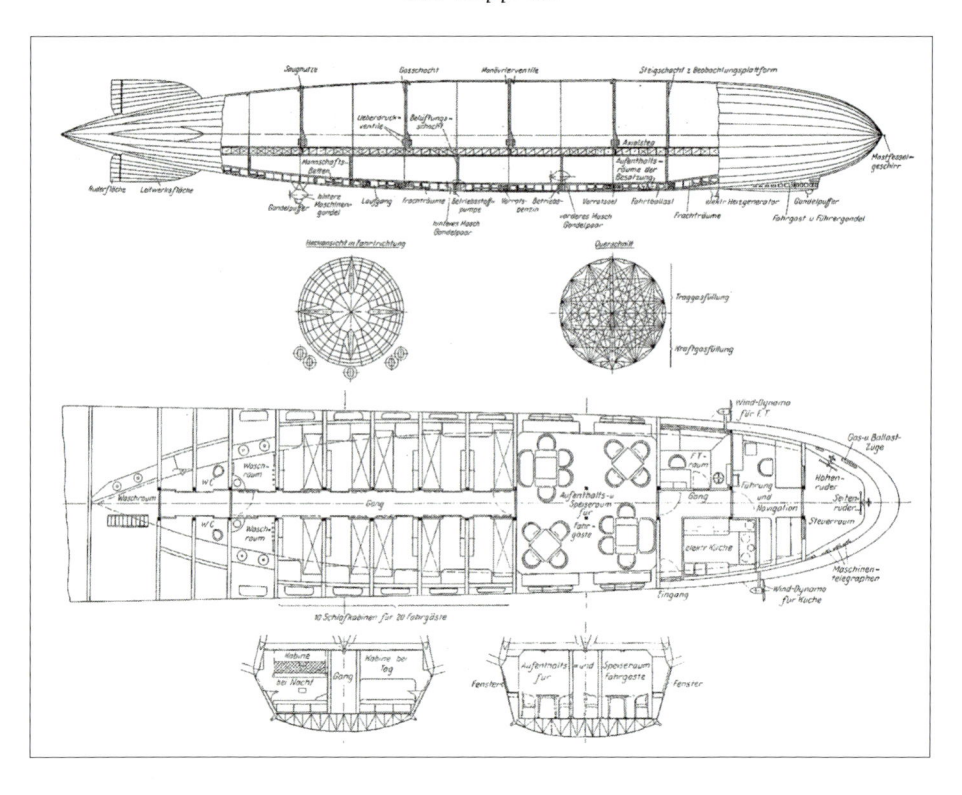

Above: A diagrammatic view of *Graf Zeppelin*.

Left: Graf Zeppelin 1934 South America schedule. *Graf Zeppelin* first flew on 18 September 1928 and was introduced to service on 11 October 1928. She was officially retired on 18 June 1937 after the *Hindenburg* disaster. Herman Göring ordered her to be scrapped in March 1940 to release aluminium for the aircraft industry.

Three views of the dining room on *Graf Zeppelin*.

Graf Zeppelin flies over the Capitol.

Graf Zeppelin entering the harbour of Rio de Janeiro, 25 May 1930.

Graf Zeppelin flying over Wembley Stadium in London during the 1930 FA Cup Final, 26 April 1930. Arsenal beat Huddersfield Town 2–0.

Up to 1935 DELAG (*Deutsche Luftschiffahrts-Aktiengesellschaft*), was the only airline using Zeppelins. It was founded on 16 November 1909 and operated Zeppelin rigid airships manufactured by the Luftschiffbau Zeppelin Corporation. Hugo Eckener, chairman of Luftschiffbau Zeppelin had the temerity to indicate a desire to run against Hitler in the 1932 presidential election. Reich Minister of Aviation Hermann Göring insisted that a new agency be created to extend Party control over LZ Group. Deutsche Zeppelin-Reederei was therefore incorporated on 22 March 1935 as a joint venture between Zeppelin Luftschiffbau, the Ministry of Aviation, and Deutsche Luft Hansa; in this manner the Nazis side-lined Eckener. This photograph is the Deutsche Zeppelin-Reederei office in Frankfurt.

80   *The Zeppelin*

Left: An oil advertisement featuring the *Graf Zeppelin*.

Below: Two advertisement posters, the first for the Hamburg-Amerika Line, the second for the new Deutsche Zeppelin-Reederei (DZR).

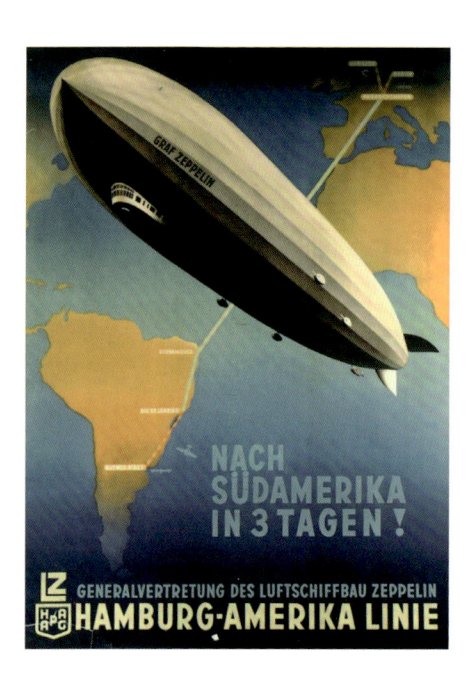

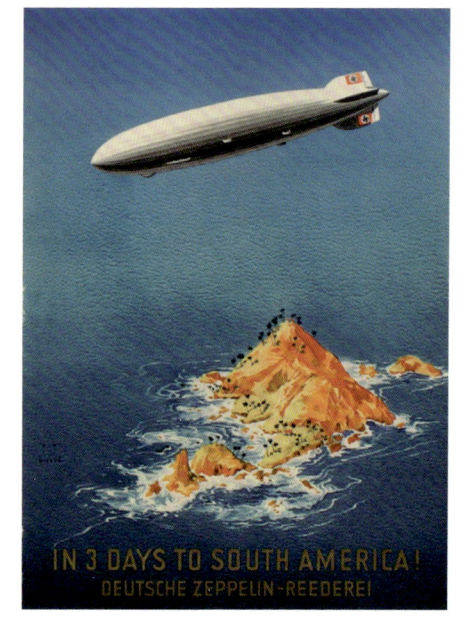

Post-War 81

The *Hindenburg* was the largest and most sophisticated airship ever built, along with her sister ship, the rarely seen *Graf Zeppelin II*. She was built for trans-Atlantic service and was seen as something of a flagship for Nazi Germany, hence the large swastika on the tail fins of the airship. These images show her passing over Lower Manhattan.

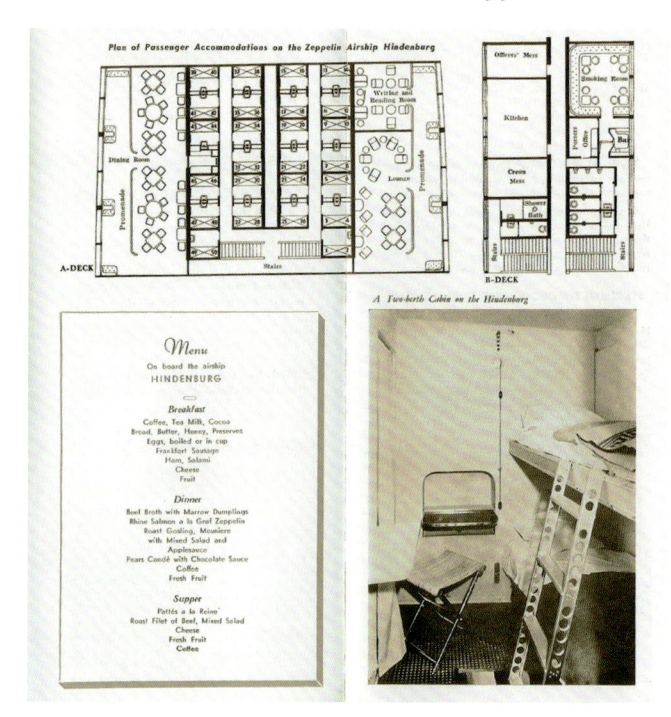

Two pages from a *Hindenburg* brochure, made interesting for the internal plan of the airship and for its typical menu.

A comic-type drawing showing the mail passenger deck on the airship. *Hindenburg* made the trips from Frankfurt to Lakehurst in an average of 63 hours and 42 minutes; the trips from Lakehurst to Frankfurt were made in 51 hours and 46 minutes; the average speed was 81 miles per hour.

Passengers disembarking on 9 May 1939 after the maiden voyage from Frankfurt to Lakehurst.

The reception after the maiden voyage. *Hindenburg* made 17 round trips across the Atlantic in 1936—its first and only full year of service—with ten trips to the United States and seven to Brazil. The first passenger trip across the North Atlantic left Frankfurt on 6 May with 56 crew and 50 passengers, arriving in Lakehurst on 9 May. The ten westward trips that season took 53 to 78 hours and eastward took 43 to 61 hours.

A brochure photograph of the dining room.

Diners during the maiden voyage. The dining room seating on *Hindenburg* accommodated all 50 passengers at the same time.

Two views of the large dining room. Cooking aboard *Hindenburg* presented a unique set of challenges. Of course, when dealing with an enormous balloon full of highly flammable gas, open flames were strictly forbidden. The galley of the ship was fully electric, from stoves to ovens. A head chef had a team of five assistants to help with all the duties of serving three meals a day to a hungry crew of 60 along with the 40 or so passengers. Roughly 440 pounds of meat and poultry were brought aboard, along with 800 eggs and 40 gallons of milk, among other supplies.

To the side of the dining room was the 'promenade', an observation area so that passengers could enjoy the sight of the world sliding by at 80 mph.

Keen observers will note that the passengers here are on the promenade on the opposite side of 'A' deck, adjacent to the lounge. The next room down off the promenade is the writing and reading room.

The lounge on *Hindenburg*. Note the large mural displaying Zeppelin routes to South America. The lounge even featured a baby grand piano that was made of aluminium and covered in yellow pigskin, weighing only 171 kg.

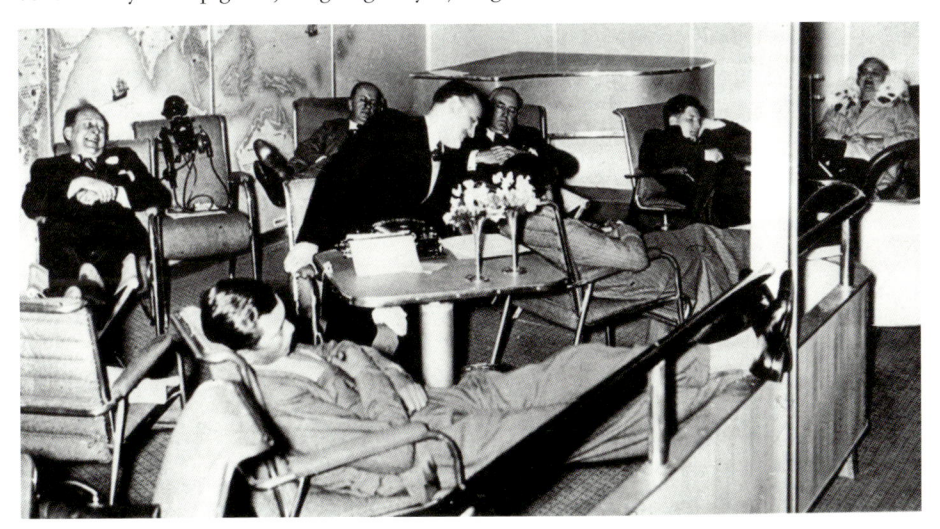

On 'B' deck, below the dining-room and lounge cabin was writing room and the comfortable smoking room. The thought of smoking with so much hydrogen above seems a little counter-intuitive, but the Germans had an answer. The room was kept under positive pressure to prevent any of the gas from entering and a single electric lighter was provided to light pipes, cigars or cigarettes.

Right: Hindenburg being manhandled. This is probably a photograph taken at Lakehurst, New Jersey in 1936. This image clearly shows the vast scale of the airship. Note the jettisoning of water ballast.

Below: Hindenburg flies over Manhattan, on 6 May 1937, with ocean liner piers below. A few hours later a disaster occurred which effectively brought the era of Zeppelins and other airships to an abrupt end.

The destruction of the *Hindenburg* at Lakehurst on 6 May 1937 remains one of the iconic images of the twentieth century. Journalists, photographers, and the public watched in horror as the airship moved in to its mooring position and then burst into flames. The cause of the explosion is unknown, but thirty-seven people died and the disaster effectively spelled the end of commercial airship travel.

These two images are exactly the same photograph, but exposed differently when printed they each show subtle differences of detail. The photograph was taken a few seconds after the spark had ignited the gas and when Hindenburg is adjacent to the Lakehurst mooring tower.

The *Hindenburg* hit the ground, stern first, the outer covering collapsing as hydrogen flames burst from the side of the airship. Ernst Lehmann, who was present on the flight deck as an observer (the captain that day was Max Pruss) died from burns received in the accident.

Forward keel provisioning hatch through which Werner Franz escaped. He was a 13-year-old cabin boy and was lucky to have a convenient exit to safety.

Adolf Fisher, an injured mechanic is transferred from Paul Kimball Hospital in Lakewood, New Jersey, to an ambulance going to another area hospital, on 7 May 1937.

Surviving members of the crew are photographed at the Naval Air station in Lakehurst, New Jersey, on 7 May 1937. Rudolf Sauter, chief engineer, is at centre wearing white cap; behind him is Heinrich Kubis, a steward; Heinrich Bauer, watch officer, is third from right wearing black cap; and 13-year-old Werner Franz, cabin boy, is centre front row. Several members of the airship's crew are wearing US Marine summer clothing furnished them to replace clothing burned from many of their bodies as they escaped from the flames.

Customs officers search through baggage items salvaged after the disaster.

Left: Fritz Deeg (1912–1990), steward on *Hindenburg*. The day after the fire, Fritz Deeg, along with fellow steward Wilhelm Balla and Chief Engineer Rudolf Sauter, had the unenviable task of trying to identify bodies of those killed in the fire.

Below: Two men inspect the twisted metal framework of *Hindenburg*.

Post-War 95

In New York City, funeral services for the 28 Germans who lost their lives in the *Hindenburg* disaster are held on the Hamburg-American pier, on 11 May 1937. About 10,000 members of German organisations lined the pier.

Germans give the Nazi salute as they stand beside the coffin of Captain Ernest A. Lehmann during a funeral service held on the Hamburg-American pier in New York City, 11 May 1937.

BIBLIOGRAPHY

The Modern Book of Flying, (Sampson Low, Marston and Co.)
Burns, I. M., *The RNAS and the Birth of the Aircraft Carrier, 1914–1918*, (Fonthill, 2014)
Carradice, P., *First World War in the Air*, (Amberley, 2012)
Carradice, P., *An Illustrated Introduction to the First World War*, (Amberley, 2014)
Credland, A. G., *The Hull Zeppelin Raids*, (Fonthill, 2014)
Deighton, L., and Schwartzman, A., *Airshipwreck*, (Book Club Associates, 1978)
Hyde, A. P., *The First Blitz*, (Leo Cooper, 2002)
Lehmann, E., *Zeppelin: The Story of Lighter-Than-Air-Craft*, (Fonthill, 2015)
Ridley-Kitts, D. G., *Military, Naval and Civil Airships Since 1873*, (The History Press, 2012)
Stephenson, C., *Zeppelins: German Airships 1900–40*, (Osprey, 2004)